COMPETITIVE PHYSICS 1

INTRODUCTION

This objective physics series provides a basic and challenging problem of physics from particular topics. It can be used to brush up ones basics and checking up the preparation level of particular topics. It is equally helpful to the traditional classes as well as competitions. It can be also taken as a revision material for any competition which includes the test of basic physics. If you want to grasp the subject before practicing these multiple choice questions, you can go through the website http://www.ncert.nic.in/ncerts/textbook/textbook.htm and down load the free copy of science books and after having command on the topic practice it. For revision purpose, important points are given at the starting of each topic.

CONTENTS

1. MOTION

SOME IMPORTANT POINTS

- When an object changes it position with time that would be motion.
- The origin is the reference point which use take to describe position of an object.
- The total path covered by an object is its direction.
- Magnitude is the numerical value of a physical quantity.
- The shortest distance covered from initial to the final position an object is called its displacement.
- The magnitude of displacement may be zero some time.
- When an object covers equal distance in equal intervals of time, it is said to be in uniform motion.
- The speed of an object is the distance covered by the object per unit time.
- The speed of an object with direction is called velocity.
- The SI unit of speed or velocity is m/s.
- The rate of change of velocity of an object per unit time is called acceleration.
- The Sin unit of acceleration is m/s^{-2}.
- The circular motion is an example of accelerated motion.
- The SI unit of distance and displacement is metres.
- When an object moving with uniform speed in distance – time graph, a straight line inclined at 45^0 with time.
- When a line parallel to time axis in a velocity time graph it is said to be in uniform motion.
- The equation of motion in which v = final velocity u initial velocity, a = acceleration, s = displacement.
- $v = u + at$, $\quad s = ut + 1/2at^2$, $\quad 2as = v^2 - u^2$.

1. MOTION

1.   An object is said to be in motion when:

   a. it changes its size   b. it changes its state

   c. it changes its position with time   d. it changes its direction

2.   To describe the position of an object we specify:

   a. a reference point   b. direction

   c. time   d. none of these

3.   The numerical value of a physical quantity is its:

   a. size   b. state

   c. magnitude   d. none of these

4.   The SI unit of distance is?

   a. meter   b. meter/second

   c. kg/second   d. km/second

5.   An object is said to be in uniform motion when it covers:

   a. equal distance in unequal interval of time.

   b. equal distance in equal interval of time.

   c. unequal distance in equal interval of time.

   d. none of these.

6. Meter per second is the SI unit of:

   a. distance   b. displacement

   c. speed and velocity   d. none of these

7.   To specify the speed of an object we only require its:

a. magnitude   b. SI unit

c. direction   d. none of these

8. The speed of an object is a:

   a. scalar   b. vector

   c. none of these   d. both a and b

9. The average speed of an object is obtained by:

   a. total distance/total time taken

   b. total displacement/ total time taken

   c. total speed/ total time taken

   d. none of these

10. An object covers 20 km in 2000 seconds and the other 30 km in 2500 seconds. What is the average speed of the object?

   a. 5m/s   b. 8m/s

   c.11.1m/s   d. none of these

11. When an object has speed with direction it shows:

   a. velocity   b. acceleration

   c. displacement   d. none of these

12. The distance covered by any object can be calculated by a device called:

   a. speedometer   b. odometer

   c. thermometer   d. none of these

13. The velocity of an object can be changed by changing:

   a. the object's magnitude   b. direction of motion

c. both a and b   d. none of these

14.   The velocity of an object remains constant with time in:

a. uniform motion   b. non uniform motion

c. both a and b   d. none of these

15.   When an object is moving with changing velocity, the object is said to be in:

a. uniform motion   b. non uniform motion

c. both a and b   d. none of these

16.   The SI unit of acceleration is:

a. m/s   b. m/s^2

c. kg/s^2 d. m

17.   When acceleration of an object is opposite to the direction of its velocity, the acceleration is taken as:

a. positive   b. negative

c. neutral   d. none of these

18.   When a body falling freely from a height it is said to be in:

a. uniformly accelerated motion   b. non uniformly accelerated

c. none of these   d. both a and b

19.   A car decreases its velocity from 90m/s to 30m/s in 10s. Find the acceleration of the car?

a. 6m/s   b.-6m/s

c. 8m/s   d. 3m/s

20.   The acceleration of an object is a :

a. scalar quantity   b. vector quantity

c. independent   d. None of these

21. Which independent quantity is taken as a reference in any graph?

a. speed   b. Time

c. distance   d. none

22. In distance time graph which term is taken along the y-axis?

a. time   b. distance

c. velocity   d. none of these

23. When the distance time graph is straight line inclined at 45 degree with time axis it shows:

a. uniform motion   b. non uniform motion

c. object at rest   d. none of these

24. When the distance time graph is straight line parallel to time axis, it shows object is in:

a. non uniform motion   b. at rest

c. uniform motion   d. none of these

25. When the speed time graph is a straight line parallel to time axis, it shows:

a. none uniform motion   b. uniform

c. rest   d. none of these

26. When a car moving with uniform acceleration shows, velocity- time graph is:

a. a straight line inclined at 45 degree with time axis

b. a straight line parallel to time axis

c. a straight line parallel to velocity axis

d. none of these

27. When a body runs in a rectangular track once, how many times it changes its direction:

a. 1

b. 4

c. 3

d. 5

28. When an object moves with a uniform speed in a circular path it shows:

a. uniform circular velocity

b. non uniform motion

c. objects at rest

d. none

29. The SI unit of deceleration is:

a. m

b. m/s

c. m/s^2

d. $-m/s^2$

30. The rate of change of velocity is:

a. acceleration

b. speed

c. distance

d. force

31. the initial velocity of a body is u with an uniform acceleration a, its distance s at any time t is given by:

a. s=v

b. $s=ut+.5at^2$

c. $s=ut+at^2$

d. none

32. a car with speed 850m/s in 1800 second, travels a distance of

a.1530000m

b.114500m

c. 14000m

d.15000m

33. A bus increases its speed from 10km/h to 20km/h in 5 second. What is its acceleration?

   a. $0.55 m/s^2$ b. $5 m/s^2$

   c. $4 m/s^2$ d. $3 m/s^2$

34. A car starts from rest increases its velocity by 20km/h, its initial velocity is :

   a. u=20   b. u=0

   c. v=0   d. none

35. When a body falls freely, what changes rapidly :

   a. distance   b. time

   c. velocity   d. force

36. Motion of moon around earth is an example of:

   a. uniform motion   b. accelerated motion

   c. non uniform motion   d. none

37. Which statement is incorrect for distance?

   a. it can be zero   b. it is a vector quantity

   c. it can be negative   d. all of the above

38. When a body shows uniform motion its acceleration is:

   a. zero   b. more than zero

   c. negative   d. none

39. Which formula shows relation between distance s, speed v, and time t?

   a. f=ma   b. a=(v-u)/t

   c. v=s/t   d. none

40. When a body is moving with a variable acceleration, the velocity time graph is:

a. curved   b. straight

c. none   d. both a and b

41. The area enclosed by velocity time graph and the time axis is the:

a. distance travelled by object   b. speed covered by object

c. none   d. both a and b

42. When final position of object coincides with initial position, the displacement will be:

a. zero   b. increases

c. at rest   d. none of these

43. Distance of an object can be:

a. less than displacement   b. not less than displacement

c. none of these   d. both a and b

44. The acceleration of object is in the direction of velocity when:

a. u is greater than v   b. V is greater than u

c. a = positive   d. both b and c

45. When odometer of an automobile shows increment in its reading it shows:

a. distance increased   b. speed increased

c. time increased   d. none

46. A graph is always plotted between:

a. Two variable quantities   b. two same quantities

c. Three variable quantities   d. none

47.   A car starts from a point A and reach to the point B. after some time car reach back to the point A. the displacement of the car is:

a. constant   b. zero

c. One   d. none

48.   The formula of acceleration of an object is:

a. $a = (v-u)/t$ b. $a = ut+.5at^2$

c. $a = v$ d. $a = v+t^2$

49.   Which is not an equation of motion?

a. $v = u+at$ b. $f= mg$

c. $s = ut+.5at^2$ d. all of these

50.   An object starts from rest increases its velocity by 20km/h in 40 second. Its acceleration will be:

a. $0.14m/s^2$ b. $2m/s^2$

c. $5 m/s^2$ d. $8m/s^2$

ANSWERS:

QUES.	ANS.	QUES.	ANS.	QUES.	ANS.	QUES.	ANS.	QUES.	ANS.
1	C	11	A	21	B	31	B	41	A
2	A	12	B	22	B	32	A	42	A
3	C	13	C	23	A	33	A	43	B
4	A	14	A	24	B	34	B	44	D
5	B	15	B	25	B	35	C	45	A
6	C	16	B	26	A	36	B	46	A
7	A	17	B	27	B	37	D	47	B
8	A	18	A	28	B	38	A	48	A
9	A	19	B	29	D	39	C	49	B
10	C	20	B	30	A	40	A	50	A

2. FORCE AND LAWS OF M OTION

SOME IMPORTANT POINTS

➤ Force is muscular effort which can change the magnitude of velocity the direction of motion the shape or size of an object.

➤ When the applied force on a object by two sides is equal that is called balanced force.

➤ When unbalanced force acting on an object brings the object in motion.

➤ First law of motion states that an object remains its of rest or motion unless an unbalanced force act on it motion.

➤ The tendency of an object to maintain its state is called inertia.

➤ The object which has a larger mass has a larger inertia.

➤ The momentum of an object is described by the product of its mass and velocity

(p=mv)

➤ The SI unit of momentum is (kgm/s^{-1})

➤ Momentum gas both direction and magnitude

➤ The second low of motion stats that the rate of change of momentum is dire city proportional to force applied on a object.

➤ Newton's second low of motion force acting on am object.

(f=ma)

➤ Newton's and third low of motion states that every action has a equal and opposite reaction

➤ Action and reacting acts on two different objects.

➤ Forces of action and reaction are equal and opposite

➤ The SI unit of force is newton or kgms^{-2}

➤ Law of conservation of momentum shows that sum of momentumta bet or collision is equal to the sum of moment a after collision.

$(m_1u_1+m_2u_2)=(m_1v_{1+}m_2v_2)$

2. FORCE AND LAWS OF MOTION

1. The SI unit of momentum is?

   a. PaS   b. NS

   c. N/S   d. kgm/S^2

2. Which is the odd one out?

   a. force   b. momentum

   c. mass   d. acceleration

3. Force is defined as:

   a. mass × velocity   b. Mass × acceleration

   c. mass/volume   d. pressure/area

4. When a force of 35 N acts on a body of mass 7 kg, the body will be accelerated at:

   a. 5 m/s^2 b. none

   c. 35 m/s^2 d. 245 m/s^2

5. A mass of 7 kg moves at 4m/s. its momentum is

   a. 28NS   b. 70 Ns

   c. 28 Kgm/s^2 d. 70 Kgm/s^2

6. When a force of 30 N acts on a body of mass 10 kg, the body will be accelerated at

   a. 30 m/s^2 b. 3 m/s^2

   c. 300 m/s^2 d. 236 m/s^2

7. Which is the odd one?

a. speed   b. density

c. momentum   d. mass

8.   The SI unit of force is

   a. Pascal   b. joule

   c. Newton   d. dyne

9.   A mass of 9 kg moves at 2 m/s. its momentum is:

   a. 18 kgm/s^2 b. 90 kg m/s^2

   c. 18 Ns   d. 90 Ns

10. An object is moving in a circular path. The force acting on the objet towards the centre of the circle is called the

   a. magnetic force   b. centripetal force

   c. gravitational force   d. none of these

11. Which of the following statement is not true?

   a. momentum = mv   b. f = ma

   c. velocity = speed/time   d. none

12. A body continues in its state of rest or uniform motion in a straight line unless acted upon by a force is a statement of

   a. Newton's first law of motion   b. Newton's second law of motion

   c. Newton's third law of motion   d. law of gravitation

13. Action and reaction always act on different bodies in opposite directions according to which laws:

   a. First law of motion   b. second law of motion

   c. Third law of motion   d. none

14. Who wrote the book named "the little balance":

    a. Galileo galilei   b. M.K Gandhi

    c. both a and b   d . None

15. Galileo galilei was born on

    a. 15th Feb. 1564   b. 16th march 1697

    c. 15th Feb. 1565   d. 15th Feb. 1678

16. An object to resist a change in its state motion or state of rest is

    a. inertia   b. first law of motion

    c. Second law of motion   d. momentum

17. What is the Si unit of mass and acceleration respectively?

    a. m and kg   b. kg and m/s^2

    c. m/s and m   d. Ns and N/s^2

18. What is the symbol of normal force?

    a. N   b. Ns

    c. m/s   d. m/s^2

19. Ship floats on the principal of which force

    a. buoyant force   b. gravitational force

    c. both a and b   d. None of these

20. Define third law of motion

    a. to every action there is an equal and opposite reaction

    b. if it is at rest it tends to remain at rest

    c. both a and b

d. none

21. a milk tanker filled up to ¾ of its height is moving with a uniform speed on sudden application of the brake, the milk in the tank would

a. move backward   b. move forward

c. be unaffected   d. rise upward

22. a mass of 6 kg moves at 2m/s. its momentum is

a. 60Ns   b. 12Ns

c. 60 kg m/s^2   d. 12kg m/s^2

23. The resistance offered by an object to an applied force is referred to as

a. inertia   b. potential

c. reaction   d. friction

24. Which of the following has more inertia?

a. stone   b. train

c. five rupees coin   d. pen

25. Which is required to keep a moving object in motion?

a. displacement   b. force

c. gravitational pull   d. none of these

26. If the set of force acting on an object are balanced, then object must be:

a. at rest   b. moving

c. accelerating   d. none

27. Nuclear force is an example of which force

a. contact force   b. non contact force

c. gravitational force   d. none

28.   What is the momentum of an object of mass m, moving with a velocity v?

a. $(mv)^2$ b. mv^2

c. .5 mv^2   d.mv

29.   A spring scale reads 20 N as it pulls a 4 kg object across a table. What is the magnitude of the force exerted by object on the spring scale?

a. 40N   b. 20N

c. 4N   d. 5N

30.   When a bus starts suddenly the bus passengers standing in the bus tend to fall backwards. This is due to:

a. inertia   b. inertia of motion

c. inertia of direction   d. none

31.   The vector sum of all balanced forces is:

a. zero   b. one

c. Two   d. three

32.   Which law states that the external force is required to change the inertia of the object?

a. Newton's first law   b. second law of motion

c. Third law   d. none

33.   Which is the example of contact force?

a. buoyant force   b. friction force

c. weak nuclear force   d. both a and b

34.   1 N is equivalent to what?

a. 0.5 dyne   b. 10^6 dyne

c. 10^5 dyne   d. 10^7 dyne

35.   At bats man hits a cricket ball which then rolls on a level ground. After covering short distance, the ball comes to rest. The ball slow down and stop because:

a. the bats man didn't hit the ball hard

b. there is a force on the ball opposing the motion

c. velocity is proportional to the force exerted on the ball

d. all of these

36.   An object of mass 5kg is moving with a velocity of 4m/s. a constant force of 20 N acts on the object from opposite direction. What will be the velocity of the object?

a. 12 m/s   b. 0 m/s

c. 15 m/s   d.16 m/s

37.   Define impulse:

a. the effect of force applied for a short duration

b. the product of force and the time duration for which the force is applied

c. both

d. none

38.   What is the SI unit of the impulse?

a. m/s   b. same as of momentum

c. m/s^2   d. none

39. Which is a type of collision?

    a. elastic collision   b. plastic collision

    c. impulse   d. all of these

40. The interaction between two or more bodies is called:

    a. momentum   b. collision

    c. mass   d. none

41. if an object of mass 9 kg starts from rest and attains a velocity of 18 m/s after 6s, then the force acting on it is:

    a. 27 N   b.108 N

    c. 3N   d. 54N

42. The rate of change of ………….is equivalent to force applied:

    a. velocity   b. momentum

    c. displacement   d. density

43. When an object is at rest on a surface, what can you say about the forces on it?

    a. they are unbalanced forces

    b. there is no any forces

    c. all the forces cancel out each other

    d. all the forces are in the same direction

44. Electromagnetic force, gravitational force, strong nuclear force, weak nuclear force:

    For the above forces: which force is strongest?

    a. electromagnetic   b. gravitational

c. strong nuclear   d. weak nuclear

45.   for the given forces: which force acts downwards on the surface?

a. weight   b. air resistance

c. friction   d. none

46.   .... forces will have no effect on the momentum of an object:

a. unbalanced force   b. balanced force

c. opposite forces   d. none

47.   How are force, mass and acceleration related?

a. $F=m/a$ b. $F=a/m$

c. $F=ma$ d. $m=Fa$

48.   When an object reaches its maximum velocity when falling through a fluid, what do we call it?

a. acceleration   b. deceleration

c. constant velocity   d. terminal velocity

49.   The direction of friction is always ……….to the direction of the object motion:

a. equal   b. opposite

c. unrelated   d. related

50.   When an object is moving fast through a fluid how does this affect the force of friction on it?

a. the forces of friction is greater   b. the forces of friction is smaller

c. the forces of friction unaffected   d. the forces of friction is same

51. A car of mass 1000kg can produce an acceleration of 8m/s^2. Calculate the force produced by the engine ignoring friction:

a. 10000 N

b. 8000 N

c. 125 N

d. 80000 N

ANSWERS:

QUES.	ANS.	QUES.	ANS.	QUES.	ANS.	QUES.	ANS.	QUES.	ANS.
1	C	11	C	21	B	31	A	41	A
2	C	12	A	22	B	32	A	42	B
3	B	13	C	23	A	33	D	43	C
4	A	14	A	24	B	34	C	44	A
5	A	15	A	25	B	35	B	45	A
6	B	16	A	26	A	36	B	46	B
7	C	17	B	27	B	37	C	47	C
8	C	18	A	28	D	38	B	48	D
9	C	19	A	29	B	39	A	49	B
10	B	20	A	30	A	40	B	50 D	
								51 B	

3. GRAVITATION

SOME IMPORTANT POINTS

➤ The body moving along the circular path is acting towards the centre the centre by a force this force is centripetal force.

➤ The force of attraction between a object to any other object in universe this force is known as gravitational force.

➤ Every object in universe attract every other object in universe.

➤ Low of gravitation the force of attraction between any two object in universe is proportional to product of their masses and inversely proportional to the square of distance between them. The force is along live joining their centre.

➤ $F = G M*m/d^2$ [force, M, m=mass of two object d=distance]
G=universal gravitation constant.
The S.I unit of G is Nm^2kg^{-2} and value of $G=6.673*10^{-11} Nm^2kg^{-2}$

➤ Gravity is a force of gravitation of earth.

➤ When an object falls towards the earth under gravity its called free fall

➤ G is the acceleration due to gravity the unit of g is m/s^2 and value of $g=9.8m/s^2$ $g=G M/d^2$ [where M=mass of earth d=distance between object & earth]

➤ The value become grater at the poles than at equator

➤ Mass of an object is measure of its inertia mass of an object is constant and does not change place to place.

➤ It is scalar quantity and its unit is kilogram (kg)

➤ The weight of an object is a force with which it is attracted to wards the earth

➤ Weight =mass*gravity
F=mg, unit of weight is Newton

➤ it is a vector quantity

➤ Weight of an object on moon is 1/6th the weight of that object on earth.

➤ The perpendicular force acting on an object is known as thrust.

➤ The unit of thrust is Newton

- The thrust per unit area is known as pressure
- The S.I unit of pressure is N/m^2 bat in hours of scientist base Pascal the S.I unit of pressure called Pascal denoted by pa
- The force acting on smalls area exerts a larges pressure
  This is the reason of why a nail has pointed tip
- All liquids and gases are fluids
- The upward force exerted by a liquid on object is known as up thrust or buoyant force
- Archimedes principle when a body is immersed fully or partially in a fluid. It experiences an upward force that is equal the weight of fluid displaced by it
- Archimedes principle used in designing ships submarines, Lactometers, hydrometer etc.
- The relative density of a substance is the ratio of its density to that of water
  Relative density= density substances /density of water

3. GRAVITATION

1. Who calculated the value of G?

    a. Newton   b. Cavendish   c. Galileo   d. Chadwick

2. The value of G:

    a. 9.8 m/s^2 b. 60673 * 10^{-11} Nm/kg^2

    c. 9.998 * 10^{-12} Nm^2 /kg^2 d. 9.998* 10^{-11} m/s^2

3. Where the radius of earth is more:

    a. pole   b. equator

    c. both a and b   d. everywhere same

4. The value of g is:

    a.6.023*10^{23} b. 6.673 * 10^{-11}

    c. 6.673 m/s^2 d. 9.8 m/s^2

5. The mass of a man is 50 kg on the earth then what will be the value of the mass of that man on moon:

    a. 7kg   b. 50/6kg

    c. 50/2kg   d. 50kg

6. SI unit of weight is:

    a. m   b. kg

    c. N   d. ton

7. Weight is a .... quantity:

    a. scalar   b. vector

    c. both a and b   d. Constant

8.	What is the radius of the earth?

	a. $5.98 * 10^{24}$				b. $1.74 * 10^6$

	c. $1.74 * 10^{24}$				d. $6.37 * 10^6$

9.	What is the mass of the moon?

	a. $1.74 * 10^{29}$				b. $7.36 * 10^{22}$

	c. $7.36 * 10^{29}$				d. $7.36 * 10^{59}$

10.	Mass of an object is 24kg. What is the weight on earth?

	a. 24/6N				b. 4N

	c. 235.2 kg				d. 235.2 N

11.	force exerted perpendicular to the surface is known as:

	a. pressure				b. thrust

	c. burst				d. buoyant force

12.	The unit of pressure is:

	a. kg/s^2			b. N/m^2

	c. N/kg^2			d. Nm^2/kg^2

13.	The SI unit of thrust is:

	a. kg				b. N

	c. Pa				d. kgm/s

14.	The SI unit of pressure is:

	a. N				b. Pa

	c. P				d. kgm/s

15.	The thrust of a unit area is known as:

a. power   b. buoyant force

c. Pascal   d. pressure

16.   What is the Si unit of density?

a. kg/cm^3 b. kg/m

c. kg/m^2 d. kg/m^3

17.   Magnitude of the buoyant force depends on.........................of the fluid displaced:

a. quantity   b. volume

c. density   d. height of the container

18.   The upward force exerted by the fluids on the object is known as:

a. thrust   b. energy

c. up thrust   d. power

19.   Lactometers and hydrometers are based on which principle's:

a. Newton's   b. Henry Cavendish

c. J.J Thomson's   d. Archimedes

20.   Lactometer is used to measure:

a. purity of water   b. purity of acid and base

c. purity of milk sample   d. all of the above

21.   What is the density of gold (in kg/m^3):

a. 19037   b. 19012

c. 19300   d. 1880

22. The relative density of a material is 20.8. The density of water is 1000 kg/m³. Then what is the density of that material (in kg/m³):

a. 2080   b. 1800

c. 20800   d. 208000

23. When we throw an object then it falls:

a. in any direction   b. upward

c. downward   d. both b and c

24. The motion of moon around earth is due to ………………force:

a. gravitational   b. nuclear

c. electromagnetic   d. muscular

25. According to the universal law of gravitation the force between object is directly proportional to the product of their:

a. speed   b. distance

c. mass   d. weight

26. The universal law of gravitation is applicable to …………………:

a. small bodies   b. big bodies

c. all of these   d. terrestrial bodies

27. Which force binds us to earth?

a. Muscular force   b. gravitational force

c. nuclear force   d. centripetal acceleration

28. When an object is in the free fall it's………of the motion do not change:

a. speed   b. velocity

c. direction   d. displacement

29. When an object is in the free fall its velocity changes due to:

  a. acceleration   b. gravitational force

  c. speed   d. direction

30. When an object is in the free fall its velocity:

  a. remains same   b. Increases

  c. decreases   d. depends on the nature of the object

31. Due to which force all objects in the universe attract each other:

  a. electromagnetic force   b. nuclear force

  c. gravitational force   d. centripetal force

32. The weight of a man on the moon is 6 N. What is its mass on the moon? Given $g=10m/s^2$:

  a. 100/ (98*36) kg   b.1/98 kg

  c. 36/98 kg   d. 3.6kg

33. The mass of an object on the moon is 10kg what is its weight on earth:

  a. 0.00098N   b. 0.98 N

  c. 9.8 * 10^-2   d. 9.8* 10^-3N

34. The density of silver is (in kg/m^3):

  a. 1000   b. 10080

  c.10800   d.11870

35. An object of mass 20 kg falls from a certain height. Calculate the force it experience:

a. 20N   b. 199N

c. 196N   d. 197N

36.   An object fall from a certain height it experiences a force of 0.966N. Then calculate its weight on moon

a. 161N   b.966N

c. 1.61 N   d.966 kg

37.   An object falls from a certain height. It experiences a force of 0.966 N forces then what will be its weight on moon?

a. $161*10^{-3}$ N   b. 0.966N

c. $161*10^{-2}$ N   d. 0.966Kg

38.   The mass of an object remains same on…….?

a. only earth   b. only moon

c. only planets   d. all over the universe

39.   The value of acceleration due to gravity is?

a. same at poles and equator   b. more at equator

c. more at pole   d. remains same every where

40.   he atmosphere is held to the earth by?

a. earth magnetic field   b. air

c. atmosphere   d. gravity

41.   When the distance between two mass is doubled then the force becomes?

a. 1/6 times   b. 4 times

c. ½ times   d. ¼ times

42. The relationship between "G" and "g" is?

    a. $G = gM/(d^2)$ b. $g = GM/(d^2)$

    c. $g = Md^2/g$ d. $g = d^2/GM$

43. Newton's Law of gravitation is applicable on?

    a. earth   b. moon

    c. all over the universe   d. atmosphere

44. The density of water is?

    a. $1100 \ kg/m^3$ b. $1000 \ kg/m^4$

    c. $1000 \ kg/m^3$ d. $1001 \ kg/m^3$

45. The mass of an object on the moon is 1000 kg. What is its mass on the earth?

    a. 1000kg   b. 1000/6 kg

    c. 6000 kg   d. 166.666 kg

46. When the masses of the two objects are doubled. The gravitational force between them becomes?

    a. ¼ times   b. 4 times

    c. 1/6 time   d. 6 times

47. "G" is?

    a. gravitational constant   b. power

    c. energy   d. gravity

48. What is the unit of "G"?

    a. Nm^2/kg^2 b. Nm^2kg^2

c. Nm/kg^2 d. N/kg^2

49.   What is the value of "G" on moon?

a. 6.673*10^{-23} b. 6.673*10^{-11}

c. 6.673*10^{-12} d. 6.689*10^{-13}

50.   Mass is?

a. scalar quantity   b. vector quantity

c. none of these   d. vector having no direction

ANSWERS

Ques.	Ans.	Ques.	Ans.	Ques.	Ans.	Ques.	Ans.	Ques.	Ans.
1	B	14	B	27	B	40	D		
2	B	15	D	28	C	41	D		
3	A	16	D	29	B	42	B		
4	D	17	C	30	B	43	C		
5	D	18	C	31	C	44	C		
6	C	19	D	32	D	45	C		
7	B	20	C	33	C	46	B		
8	D	21	C	34	B	47	A		
9	B	22	C	35	B	48	A		
10	D	23	C	36	B	49	B		
11	B	24	A	37	A	50	A		
12	B	25	C	38	D				
13	B	26	C	39	C				

4. WORK AND ENERGY

SOME IMPORTANT POINTS

➤ When some force applied on a object and object must be displaced when we say that work is done.

➤ The unit of work is Joule.

➤ When 1N force applied on a object and object displaced 1m then we say that work done is 1J.

➤ 1J=1N*1m.

➤ The capacity to can do a work is known as energy. The unit of energy is joule also.

➤ There are many types of energy like kinetic energy, potential energy, mechanical energy, solar energy, nuclear energy etc.

➤ Object posses energy due to motion it said to be energy.

➤ Increased in speed kinetic energy is also increased.

➤ Objects possess energy due to change its height, shape, size it said to be potential energy.

➤ When work done against the gravity the energy possessed gravitational potential energy.

➤ Increase in height gravitational potential energy.

➤ Law of conservation of energy states the energy gets transformed one form to another; it can neither be created nor be destroyed.

➤ The total energy before the transformation and after transformation remains same it is valid in all situation.

➤ Potential energy+ Kinetic energy=Constant.

➤ The rate of doing work is known is power.

➤ The unit of power is watt and 1watt=1J/1sec.

➤ 1KW=1000W

➤ 1 horse power=746W.

➤ The energy used in 1hour of rate of 1KW known as1KWh.

➤ 1KWh=$3.6*10^6$ Joule.

➤ Work=Force*displacement

- W=F*S
- Kinetic energy=1/2*mass*(velocity)^2.
- E_k=1/2mv^2
- E_k=P^2/2m   (P=momentum, m=mass)
- Potential energy=mass*velocity*acceleration due to gravity
- E_p=m*g*h=mgh
- Power=Work/time
- 1unit=1KWh.

4. WORK AND ENERGY

1. What two conditions need to be satisfied for work to be done?

    a. a force should act on object and the object must be displaced.

    b. a force should act on object and the object should not be displaced.

    c. a force should act on object and the shape of the object must be changed.

    d. both a & c.

2. What is the symbol of the work done?

    a. F   b. N   c. W   d. G

3. What is not the formula of the work done?

    a. F.v   b. F.s   c. ½ mv²   d. mgh

4. What is the SI unit of work?

    a. watt   b. Nm   c. N   d. N/s

5. A force of 7N is acting on an object and it is displaced through 20cm in direction of the force. Then the work done is?

    a. 20j   b. 24j   c. 19j   d. 21j

6. A force of 9N is acting on an object and it is displaced through 3m in direction of the force. Then the work done is?

    a. 18j   b. 7.2j   c. 1.8j   d. 1.6j

7. Work done by a force can be?

    a. negative   b. positive   c. both a and b   d. none of these

8. A man lifts a stone of 3kg from the ground and put it on his head 1.3m above the ground. Determine the work done by him on the stone.

a. 39j   b. 40j   c. 41j   d. 37j

9.   Which is the biggest source of energy for us?

a. earth   b. moon   c. sun   d. none of these

10.   What is the SI unit of energy?

a. Joule   b. Newton   c. Watt   d. Kilogram

11.   1 Kilo Joule (Kj) =?

a. 1000j   b. 746j   c. 900j   d. 786j

12.   Kinetic energy + potential energy = ... energy.

a. Mechanical   b. nuclear   c. electrical   d. chemical

13.   Object in motion posses which type of energy?

a. potential energy   b. light energy

c. chemical energy   d.kinetic energy

14.   When kinetic energy increases?

a. when potential energy increases   b. when potential energy decreases

c. when mechanical energy decreases   c. none of these

15.   What is the formula of kinetic energy?

a. ma   b. $1/2mv^2$ c. mgh   d. F.s

16.   An object of mass 32kg is moving with a uniform velocity of 5m/s. what is the kinetic energy possessed by the object.

a. 160j   b. 1300j   c. 400j   d. 650j

17.   a object of mass 10kg moving with a uniform velocity of 2m/s. what is the kinetic energy possessed by the object?

a. 20j   b. 30j   c. 40j   d.10j

18.   An object of mass 20kg moving with uniform velocity of 3m/s. what is the kinetic energy possessed by the object?

a. 126j   b. 546j   c. 710j   d. none of these

19.   What is the work to be done to increase the velocity of a bus from 18km/h to 35km/h if the mass of the bus is 4500kg?

a. 168640j   b. 168750j   c. 167750j   d. none of these

20.   What is the work to be done to increase the velocity of car from 9km/h to 54km/h if mass of car is 2500kg?

a. 228750j   b. 218650j   c. 218750j   d. none of these

21.   What is the work to be done to increase the velocity of a car from 20km/h to 60 km/h if mass of the car is 1800kg?

a. 22222.22j   b. 222222.22j

c. 2222222.22j   d. none of these

22.   What is the formula of potential energy?

a. F.s   b. mgh   c. $1/2mv^2$ d. w/t

23.   Find the energy possessed by an object of mass 20kg when it is at a height of 8m above the ground. Given g = 9.8 ms^{-2}.

a. 1578j   b. 1567j   c. 1558j   d. 1568j

24.   Find the energy possessed by an object of mass 100kgif it is at height of 3m above the ground. Given g = 9.8 ms^{-2}.

a. 2740j   b. 2940j   c. 2840j   d. 2640j

25. An object of mass 20 kg is at a certain height above the ground if the potential energy of the object is 1960j. Find the height at which the object is with respect to ground.

a. 8m   b. 12m   c. 10m   d. 11m

26. An object of mass 15 kg is at a certain height above the ground if the potential energy of the object is 300j. Find the height at which the object is with respect to ground.

a. 2m   b. 3m   c. 7m   d. 4m

27. When an object falls on ground from certain height "h". Its potential energy will be changes into ....energy.

a. chemical   b. electrical   c. mechanical   d. kinetic

28. Rate of doing work is known as?

a. power   b. kinetic energy

c. acceleration   d. none of these

29. What is the SI unit of power/

a. Joule   b. Newton   c. Watt   d. None of these

30. 1 Watt = ...?

a.   1 Joule/s   b. 1 N/s   c. 1 Hz/s   d. 1 m/s

31. 1 Horse Power = ...watt?

a. 756   b. 786   c. 746   d. None of these

32. A girl having weight 200 N climb up a rope through a height of 4 m. Girl takes 10s to accomplish this task. What is the power of this task?

a. 800W   b. 80W   c. 50W   d. 70W

33. A boy of weight 150N runs up a staircase of 60 steps in 10s. if the height of each step is 20 cm. find the power?

   a. 180W   b. 1800W   c. 280W   d. 780W

34. 1 KWh =..............J.

   a. $3.6*10^5$ b. $36*10^3$ c. 360000   d. None of these

35. An electric fan f 150 watt is used for 5h per day. Calculate the "units" of energy consumed in one day by the fan.

   a. 75   b. 0.75   c. 8.5   d. 0.7

36. An electric bulb of 20W and two fan of 15W and television of 250W are used for 7h per day. Calculate the total units of energy consumed in 7 days by these appliances.

   a. 20.78   b. 20.67   c. 20.14   d. None of these

37. Work done on an object by a force would be zero if the displacement of the object is...........................?

   a. zero   b. more than zero

   c. work can't be zero   d. None of these

38. Which has the same unit as of work?

   a. power   b. force   c. energy   d. None of these

39. Which can neither be created nor be destroyed?

   a. energy   b. mass   c. both a and b   d. none of these

40. The energy used in 1 hour at the rate of 1KW is called.....................?

   a. 1kwh   b. 1kw   c. 1 h   d. none of these

41. The potential energy of a free falling object decreases and kinetic energy.......?

a. also decreases   b. increases   c. remains same   d. none of these

42. A body is under action of the equal and opposite forces. The work done by the body is?

     a. 49J   b. 0   c. -49J   d. 35J

43. If the mass of the body becomes 4 times, its kinetic energy?

     a. increases 4 times   b. gets doubled

     c. remains same   d. increases 8 times

44. If the speed of the train is increased by 4 times. Its kinetic energy will be increased b y?

     a. 4 times   b. 12 times   c. 20 times   d. 16 times

45. Watt is the unit of?

     a. rate of doing work   b. rate of doing power

     c. rate of change of velocity   d. None of these

46. Two bodies of equal mass are kept at height of h and 2.5h, respectively. The ratio of their potential energy is?

     a. 5:4   b. 5:2   c. 4:2   d. 2:5

47. kgm^2s^{-3} is associated with?

     a. energy   b. momentum   c. force   d. power

48. 1MJ = ...?

     a. 10^6J   b. 10^4J   c. 10^5J   d. 10^3J

49. A body is thrown up with kinetic energy of 25J, if it attains maximum height 10m. Find the mass of the body. Given g = 10m/s².

     a. 0.2 kg   b. 0.7 kg   c. 0.15kg   d. 0.25kg

50. How much time does it take to perform 250j of work at a rate of 25W?

a. 25s   b. 20s   c. 50s   d. 10s

Answers:

Que.	Ans.	Que.	Ans.	Que.	Ans.	Que.	Ans.	Que.	Ans.
1	D	11	A	21	B	31	C	41	B
2	C	12	A	22	B	32	B	42	B
3	A	13	D	23	D	33	A	43	A
4	B	14	B	24	B	34	D	44	D
5	D	15	B	25	C	35	B	45	A
6	C	16	C	26	A	36	D	46	D
7	C	17	A	27	D	37	A	47	D
8	A	18	A	28	A	38	C	48	A
9	C	19	B	29	C	39	C	49	D
10	A	20	C	30	A	40	A	50	D

SOUND

SOME IMPORTANT POINTS

- Sound produced by vibration of object.
- Sound is the form of mechanical energy.
- Propagation of sound waves as series of compression and rarefaction called longitudinal wave.
- Change in density from maximum value to minimum value.
- The number of complete oscillation in density per second of a sound wave called frequency.
- Frequency is denoted by μ.
- S.I unit of frequency is hertz (Hz).
- Distance between two consecutive compressions and two consecutive rarefactions called wavelength. It is denoted by λ (lemda).
- Speed of sound V = distance/time, $\lambda/t = v = \lambda/t$.
- Frequency is directional proportional to pitchnes.
- Amplitude is directly proportional to loudness.
- Speed of sound in solid > Speed of sound in liquid >Speed of sound in gases.
- Speed of sound increases with increasing in temperature.
- Reflection of sound: Angle of reflection is equal to angle of incidence $\rightarrow \angle r = \angle i$ reflected wave, incident wave and normal all lies in same plane.
- Repetition of sound by reflection from an absolute called echo.
- Human ears can hear only 20Hz to 20 KHz of frequency.
- The longitudinal wave's frequencies below 20Hz or called infrasonic, longitudinal waves frequencies lie 20 KHz called ultrasonic.
- SONAR is sound navigation and ranging used to measure distance in sea.

5. SOUND

1.   How does sound is produced?

   a. mass   b. vibration c. weight   d. energy

2.   Sound is a form of?

   a. energy   b. motion   c. power   d. force

3.   For transferring sound energy we need:

a. propagation   b. vibration c. medium   d. all of these

4.   The sound of human voice is produced due to vibration of:

       a. Brain   b. Neck   c. Vocal chords   d. teeth

5.   Sound can't travel through:

       a. solid   b. liquid   c. gas   d. vacuum

6.   A particle of the medium in contact with the vibrating object is first displaced from its:

       a. equilibrium position   b. Rest position

       c. Far position   d. all of these

7.   After displacing the adjacent particle the first particle comes back to its original position in a longitudinal wave:

       a. True   b. false   c. depends on temperature   d. None of these

8.   Sound waves are characterized by the motion of particle in the medium and it is called:

       a. Mechanical wave   b. Acoustic wave

       c. Electromagnetic wave   d. None of these

9.   In which matter sound travels fast?

       a. liquid   b. gas   c. jelly   d. solid

10.   When vibrating objects move forward, it pushes the air in front of it creating a region of high pressure, which is called?

       a. compression   b. rarefaction   c. pushing d. all of these

11.   Can we hear sound in vacuum?

       a. yes   b. No   c. sometimes   d. depends on speed of sound

12. Sound waves are:

a. longitudinal   b. acoustic   c. transverse   d. None of these

13. In the wave the individual particle of medium moves in a direction parallel to the direction of motion is known as:

a. Propagation   b. Wave front   c. Vibrating object   d. all of these

14. What is the SI unit of Wavelength?

a. meter   b. ampere   c. Newton   d. joule

15. In a transverse wave particle oscillate along the line of:

a. wave propagation   b. Wave front

c. perpendicular to the motion   d. none of these

16. In SONAR we use?

a. ultrasonic wave   b. infrasonic wave

c. radio wave   d. audible sound

17. Sound travels fast in which season?

a. summer   b. winter

c. rainy   d. spring

18. When we change a feeble sound to loud sound we increase its?

a. frequency   b. amplitude

c. velocity   d. wavelength

19. Melody of sound depends on:

a. frequency   b. amplitude

c. velocity   d. wavelength

20. Earthquake produces which type of sound before the main shock wave begins:

    a. ultrasound   b. infrasound

    c. audible sound   d. none of these

21. Infrasound can be heard by:

    a. dog   b. bat

    c. rat   d. rhinoceros

22. Symbol of wavelength is:

    a. σ b. λ c. v   d. π

23. The frequency of the sound wave is represented by:

    a. v   b. λ c. σ d. π

24. The wave propagation vibrates in a direction perpendicular to the direction of propagation of wave:

    a. transverse wave motion.   b. longitudinal wave.

    c. simple wave   d. none of these

25. Speed of sound is denoted by:

    a. V   b. λ c. σ d. π

26. Speed of sound is equal to :

    a. (distance/time taken) = λ/T b. P = MV

    c. P = F/A   d. None of these

27. Choose the characteristics of sound waves:

    a. frequency   b. pressure

c. explosion   d. rarefaction

28. Sound pollution is measured in:

    a. decibels   b. amplitude

    c. velocity   d. wavelength

29. Wave has frequency of 50Hz what is its time period in second?

    a. 0.02   b. 2   c. 0.0005   d. none of these

30. The loudness or softness is determined basically of its:

    a. frequency   b. amplitude   c. Velocity   d. none of these

31. Full form of SONAR is:

    a. sound navigation and rarefaction

    b. sound navigation and rairing

    c. sound navigation and ranging

    d. none of these

32. Note is a sound of:

    a. mixture of several frequencies

    b. mixture of two frequencies only

    c. a single frequency

    d. always unpleasant to listen

33. Sound travels in air if:

    a. particle of medium travel from one place to another

    b. there is no moisture in the atmosphere

    c. disturbance

d. all of these

34. When we change feeble sound to loud sound, we increases its:

    a. frequency   b. amplitude   c. velocity   d. wavelength

35. The half wavelength is:

    a. $\lambda/2$ b. 2λ c. 2.5λ d. λ^2

36. A SONAR device on submarine sends out a signal and receives an echo 5s later. Calculate the speed (in m/s) of sound in water if the distance of the object from the submarine is 3625m?

    a. 1456   b. 1450   c. 2036   d. 1540

37. Ultrasound is used in:

    a. cleaning   b. in medical

    c. both a and b   d. none of these

38. SI unit of frequency is:

    a. joule   b. Newton   c. Pascal   d. Hz

39. The frequency of sound is 100Hz. How many times does it vibrate in a minute?

    a. 600   b. 6000   c. 60   d. 60000

40. Speed of wave is :

    a. F/A   b. λ/T c. mc^2 d. none of these

41. A sound of single frequency is:

    a. tone   b. note   c. Infrasonic sound   d. none of these

42. The sensation of sound persists in our brain for about:

    a. 0.2s   b. 0.1s   c. 1s   d. 0.02s

43. Echo is a:

    a. once reflection of sound   b. multiple reflection of sound

    c. multiple reflection   d. none of these

44. The repeated reflection that results in their persistence of sound is called:

    a. Refraction   b. reverberation

    c. both a and b   d. none of these

45. Uses of multiple reflection of sound is:

    a. cleaning   b. stethoscope   c. microphones   d. both c and d

46. The number of complete oscillation per unit time is called:

    a. amplitude   b. ultrassound   c. frequency   d. none of these

47. The outer ear is called:

    a. cochlea   b. pinna   c. anvil   d. stirrup

48. Bat produces:

    a. infrasonic sound   b. ultrasonic sound

    c. beam of sound   d. None of these

49. The amount of sound energy passing each second through unit area is called:

    a. density of sound   b. amplitude

    c. intensity of sound   d. all of these

50. Ultrasound may be employed to break the small:

    a. stones   b. light particle   c. atoms   d. both a and c

ANSWERS:

QUE.	ANS.	QUE.	ANS.	QUE.	ANS.	QUE.	ANS.	QUE.	ANS.
1	B	11	B	21	D	31	C	41	B
2	A	12	A	22	B	32	C	42	B
3	C	13	A	23	A	33	C	43	C
4	C	14	A	24	A	34	B	44	B
5	D	15	C	25	A	35	A	45	D
6	A	16	A	26	A	36	B	46	C
7	A	17	B	27	D	37	C	47	B
8	A	18	B	28	A	38	D	48	B
9	D	19	A	29	A	39	A	49	C
10	A	20	B	30	B	40	B	50	A

NOTES